Little Pebble™

Simple Machines
Pulleys

by Martha E. H. Rustad

CAPSTONE PRESS
a capstone imprint

Little Pebble is published by Capstone Press,
1710 Roe Crest Drive, North Mankato, Minnesota 56003
www.mycapstone.com

Library of Congress Cataloging-in-Publication Data
Names: Rustad, Martha E. H. (Martha Elizabeth Hillman), 1975– author.
Title: Pulleys / by Martha E.H. Rustad.
Description: North Mankato, Minnesota : Capstone Press, 2018. | Series:
    Little pebble. Simple machines | Audience: Ages 4–7.
Identifiers: LCCN 2017031576 (print) | LCCN 2017035877 (ebook) |
    ISBN 9781543500868 (eBook PDF) | ISBN 9781543500745 (hardcover) |
    ISBN 9781543500806 (paperback)
Subjects: LCSH: Pulleys—Juvenile literature.
Classification: LCC TJ1103 (ebook) | LCC TJ1103 .R87 2018 (print) | DDC
    621.8—dc23
LC record available at https://lccn.loc.gov/2017031576

Editorial Credits
Marissa Kirkman, editor; Kyle Grentz (cover) and Charmaine Whitman (interior), designers;
Jo Miller, media researcher; Katy LaVigne, production specialist

Image Credits
Alamy: VStock, 15; Capstone Studio: Karon Dubke, 11, 15 (inset), 17; Glow Images: Cultura RF/Henn Photography, 21; Shutterstock: BigIndianFootage, 13, bogdanhoda, 7, Carlos Caetano, cover, 1, Dutourdumonde Photography, 17 (inset), Feylite, 5, jayk67, 9, Lisa S., 19, Maliflower73, 19 (inset), Mathisa, 6, Monkey Business Images, 20, spetenfia, 22

Design Elements
Capstone

Printed and bound in the USA.
010766S18

Table of Contents

Help with Work

Work is hard!

We need help.

Use a simple machine.

These tools help us work.

pulley

A pulley helps us move a load.

It lifts up heavy things.

pulley

load

Parts

See the rope.

It goes on a wheel.

Pull down.

The wheel turns.

The other side goes up.

wheel

Put a load on one side.

Pull.

Lift the load!

load

Everyday Tools

A flagpole has a pulley.

Pull!

The flag rises.

A bike has a pulley.

Pedal!

The bike moves.

Blinds use pulleys.

Pull!

The blinds go up.

We use a simple machine.

It makes work easier and fun.

Glossary

blinds—a covering for a window; a pulley helps to open and close blinds

load—an object that you want to move or lift

simple machine—a tool that makes it easier to move something

wheel—a round disc

work—a job that must be done

Read More

La Machia, Dawn. *Pulleys at Work.* Zoom in on Simple Machines. New York: Enslow, 2016.

Schuh, Mari. *Raising a Bag of Toys: Pulley vs. Inclined Plane.* First Step Nonfiction: Simple Machines to the Rescue. Minneapolis: Lerner Publications, 2016.

Weakland, Mark. *Fred Flintstone's Adventures with Pulleys: Work Smarter, Not Harder.* Flintstones Explain Simple Machines. North Mankato, Minn.: Capstone Press, 2016.

Internet Sites

Use FactHound to find Internet sites related to this book.

Visit www.facthound.com

Just type in 9781543500745 and go.

Super-cool stuff!

Check out projects, games and lots more at
www.capstonekids.com

Critical Thinking Questions

1. What two parts does a pulley have?

2. What happens when you pull on one side of a pulley's rope?

3. What types of pulleys have you used?

Index